Animals Live Here

Life in the Rain Forest

By Connor Stratton

level 2 little blue readers

www.littlebluehousebooks.com

Little Blue House is distributed by North Star Editions:
sales@northstareditions.com | 888-417-0195

Produced for Little Blue House by Red Line Editorial.

Photographs ©: Dmitry Rukhlenko/Shutterstock Images, cover; Ramdan_Nain/iStockphoto, 4 (top); Dave Primov/Shutterstock Images, 4 (bottom); undefined undefined/iStockphoto, 7; Dirk Freder/iStockphoto, 9, 24 (bottom right); Mark Kostich/iStockphoto, 11; ABDesign/iStockphoto, 12, 24 (bottom left); Mats B/Shutterstock Images, 14–15, 24 (top right); Natali Glado/Shutterstock Images, 16–17; Oriol Querol/Shutterstock Images, 18, 24 (top left); Gary Gray/iStockphoto, 21; Kung_Mangkorn/iStockphoto, 22–23

Library of Congress Control Number: 2019908642

ISBN
978-1-64619-024-9 (hardcover)
978-1-64619-063-8 (paperback)
978-1-64619-102-4 (ebook pdf)
978-1-64619-141-3 (hosted ebook)

Printed in the United States of America
Mankato, MN
012020

About the Author

Connor Stratton enjoys learning about rain forests, spotting new animals, and writing books for children. He lives in Minnesota.

Table of Contents

tree

Rain Forest Animals

The rain forest has many trees, and it is wet. Many different animals live here.

Orangutans live in the rain forest.

They have fur.

They can hang from tree branches.

branch
orangutan
fur

Macaws live in the
rain forest.
These parrots have bright
and colorful feathers.
They can fly above
the trees.

macaw
feather

Pythons live in the
rain forest.
These snakes have scales.
They can climb up trees.

python
scale

frog

Many Colors

Frogs live in the rain forest.
Some frogs are green,
orange, and blue.
They have big red eyes.

Jaguars live in the rain forest.

They have yellow fur with black spots.

They hunt for their food.

jaguar

Chameleons live in the rain forest.
They can change colors.

chameleon

elephant
trunk

Big and Small

Elephants live in the rain forest. They are the biggest animals on land. They have long trunks.

Butterflies live in the rain forest.

These insects are small, light, and have colorful wings.

wing
butterfly

Sloths live in the rain forest. They move slowly. They can hang from branches.

sloth

Glossary

elephants

jaguar

frog

macaw

Index